Regenerative Engineering Society

AA008686

Held at the 2023 AIChE Annual Meeting

Orlando, Florida, USA
5-10 November 2023

ISBN: 978-1-7138-9319-6

Printed from e-media with permission by:

Curran Associates, Inc.
57 Morehouse Lane
Red Hook, NY 12571

proceedings.com

Some format issues inherent in the e-media version may also appear in this print version.

Printed with permission by Curran Associates, Inc. (2024)

For permission requests, please contact AIChE
at the address below.

AIChE
120 Wall Street, FL 23
New York, NY 10005-4020

Phone: (800) 242-4363
Fax: (203) 775-5177

www.aiche.org

Additional copies of this publication are available from:

Curran Associates, Inc.
57 Morehouse Lane
Red Hook, NY 12571 USA
Phone: 845-758-0400
Fax: 845-758-2633
Email: curran@proceedings.com
Web: www.proceedings.com

TABLE OF CONTENTS

10a Delivery of Biologics for Muscle Healing after Rotator Cuff Injury

Johnna Temenoff

After rotator cuff tendon tear, marked degeneration of the attached muscle is apparent clinically, with both fibrous and fatty infiltration of the tissue. Our laboratory is working on delivery strategies for biologics, including proteins and cells, that might slow or reverse this degeneration. In particular, our laboratory has focused on "jump-starting" host regenerative processes through use of glycosaminoglycan (GAG)-based biomaterials for release of cytokines to promote the recruitment of pro-healing cell populations, such as pro-resolving macrophages, and mesenchymal stromal cells (MSCs), into the muscle. This presentation will summarize our work in this area, including the use of both bone marrow mobilizing agents and localized delivery strategies to alter the cellular milieu in the associated muscle after rotator cuff tendon injury.

10b Bioengineered perfused human brain microvasculature for brain tumor and neural stem cell research

Guohao Dai

Blood vessels play an increasingly important role in most human tissue and organ systems. Importantly, vascular niche was found to be a key element of many stem cell environments such as neural stem cells and cancer stem cells. Vascular cells not only form conduits to deliver nutrient and oxygen, but also provide instructive signals to control stem cell self-renewal and differentiation, therefore, is critical for tissue regeneration. The mission of Vascular Bioengineering Laboratory is to integrate bioengineering approaches with stem cells and vascular biology to understand blood vessel regeneration and vascular disease processes, and to develop novel therapeutic modalities to treat vascular-related disorders such as cardiovascular, neurovascular and cancer. Toward this goal, our lab has developed the method to bioengineer human brain microvascular network consists of human brain endothelial cells, pericytes and astrocytes. We have shown that interstitial flow promotes lumen formation, interconnectivity and astrocytes association of the bioengineered vasculature and blood brain barrier functions. Furthermore, perfused bioengineered vasculature enhances neural stem cell self-renewal and neuronal differentiation and maturation. In this talk, I will present research projects on the bioengineer 3D human brain vascular network and its application in neural stem cell and brain tumor research.

10c Tailoring Intelligent Protein Engineered Biomaterials for Regenerative Medicine

Jin Kim Montclare

Through centuries of evolution, nature has developed biopolymers capable of folding and assembling into discrete structures with a functional consequence. Inspired by this, our lab focuses on engineering "intelligent" protein biomaterials with entirely new properties and function. We employ synthetic and chemical biology to construct our materials and endow them with stimuli-responsiveness. In particular, our lab has fabricated protein-derived nanomaterials: helix-elastin block polymers and coiled-coil fibers. We investigate the fundamental self-assembly and molecular recognition capabilities of these systems. Central to this work is the integration of stimuli-responsive domains through rational and computational design. More importantly, we are able to harness these structure as well as others to interface with small molecules, protein therapeutics and cells for applications in regenerative medicine.

10d Engineer Off-the-Shelf CAR-Neutrophils for Targeted Chemoimmunotherapy Against Glioma

Xiaoping Bao

Introduction: Glioblastoma (GBM), the most common type of primary brain tumor, is characterized by high mortality rate, short lifespan, and poor prognosis with a high tendency of recurrence. Functional therapeutics, including PRMT5 inhibitors, radiosensitizers, and emerging chimeric antigen receptor (CAR)-T immunotherapy, have been developed to treat GBM. However, the existence of physiological blood-brain barrier (BBB) or blood-brain-tumor barrier has impeded the efficient delivery of such promising therapeutics into the brain and limited their therapeutic efficacy. Given the native ability of neutrophils to cross BBB and penetrate the brain parenchyma [1], here we tested the therapeutic concept that neutrophils could be engineered with synthetic CARs to efficiently cross BBB and specifically target GBM for the first time. Primary neutrophils are short-lived and resistant to genetic modification. Therefore, we genetically engineered human pluripotent stem cells with different CAR constructs and differentiated them into functional CAR-neutrophils [2]. We also tested the therapeutic potential of CAR-neutrophils in delivering chemotherapeutic drugs to GBM across BBB as a dual chemoimmunotherapy.

Methods: Human pluripotent stem cell line H9 was used in this study for neutrophil and natural killer (NK) cell differentiation. CRISPR/Cas9 was used to knock different synthetic chlorotoxin (CLTX) CARs into the AAVS1 safe harbor locus of H9 for making clonal hPSC lines that express stable and homogenous CARs. Biodegradable mesoporous organic silica nanoparticles were synthesized and employed to load hypoxia-activated prodrug tirapazamine (TPZ) or clinical chemo-drug temozolomide and JNJ64619187 into hPSC-derived CAR-neutrophils. In situ xenograft mouse models of glioblastoma were constructed to evaluate antitumor activities of CAR-neutrophils and CAR-NK cells.

Results: Stable CAR-expressing hPSC lines were successfully established via CRISPR/Cas9-mediated homologous recombination (Fig. 1a). A new chemically-defined protocol was implemented to induce CAR-hPSCs into CAR-neutrophils (Fig. 1b). As compared to CAR-NK cells, systemically administered hPSC-derived CLTX CAR-neutrophils significantly reduced tumor burden in xenograft mouse models and extended their lifespan (Fig. 1c-e), suggesting the superior abilities of neutrophils in crossing BBB and penetrating GBM xenograft in mice. CAR-neutrophils could also be loaded with hypoxia-activated prodrug TPZ in silica nanoparticles with rough surfaces (R-SiO2-TPZ) and specifically delivered nanodrugs into the brain, as compared to free nanodrugs (Fig. 1f). This dual chemoimmunotherapy further inhibited tumor growth and extended lifespan in xenograft mice.

Conclusions: We demonstrated the successful generation of CAR-neutrophils from hPSCs and showed their antitumor activities against GBM in xenograft models. hPSC-derived CAR-neutrophils could also be used as an effective drug carrier, which served as a new dual chemoimmunotherapy. CAR neutrophil-mediated drug delivery may provide an effective and universal strategy for specific targeting of solid tumors.

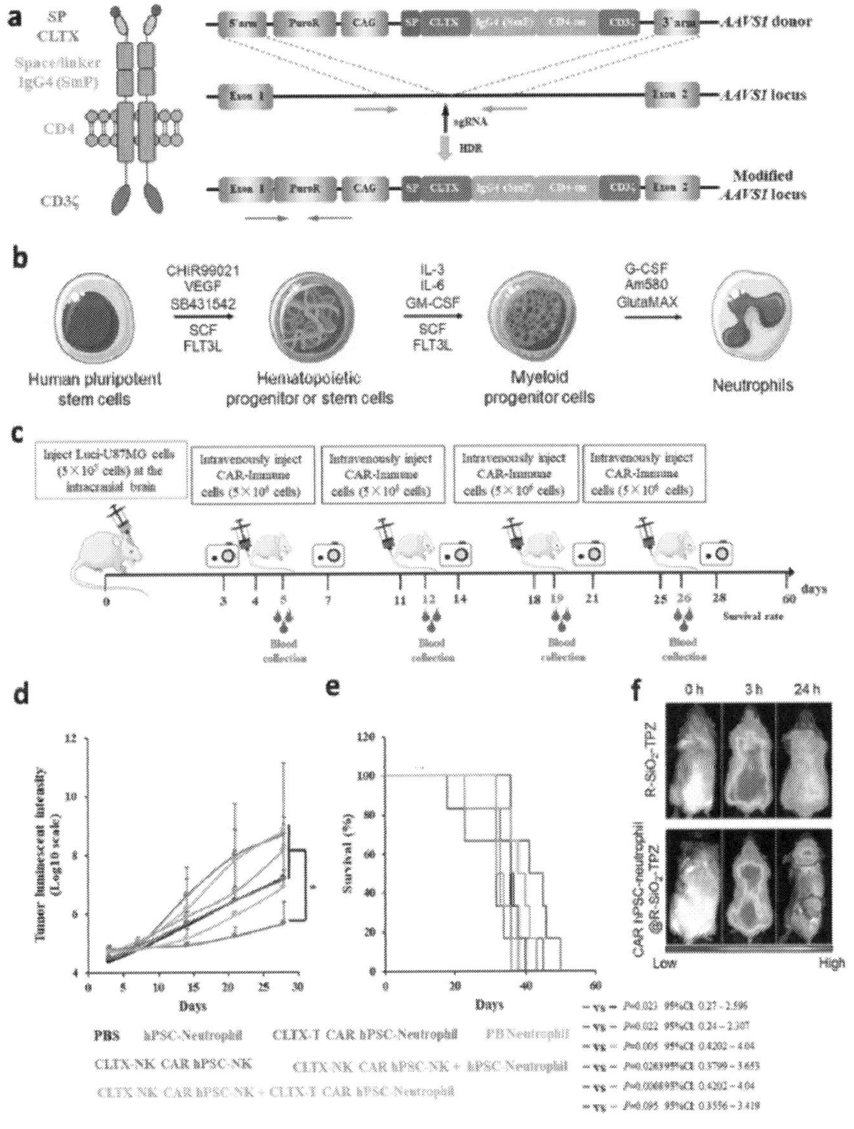

Figure 1. a) Schematic of chlorotoxin (CLTX) CAR construct and targeted knock-in strategy at the AAVS1 safe harbor locus of human pluripotent stem cells (hPSCs). b) Schematic of optimized neutrophil differentiation from hPSCs under chemically-defined conditions. c) Schematic of intravenous injection of CAR-neutrophils and/or CAR NK cells for in vivo study. d) Time-dependent tumor burden was determined and quantified by bioluminescent imaging. e) Kaplan-Meier curve demonstrating survival of experimental mice. f) Time-dependent biodistribution of Cy5+ neutrophils and nanodrugs in whole body was determined by fluorescence imaging at the indicated hours.

Conclusions: We demonstrated the successful generation of CAR-neutrophils from hPSCs and showed their antitumor activities against GBM in xenograft models. hPSC-derived CAR-neutrophils could also be used as an effective drug carrier, which served as a new dual chemoimmunotherapy. CAR neutrophil-mediated drug delivery may provide an effective and universal strategy for specific targeting of solid tumors.

10e Immunobioengineering of a Three-Dimensional Polymeric Scaffold-Based Implantable Thymic Organoid

Manpreet Bariana, Elena Cassella, Shaina Anuncio, Jakub Erben, Andrea Tuckett, Michael Poulos, Jason Butler, Johannes Zakrzewski

The thymus is the organ of T cell development and is characterized by a highly specialized microenvironment that induces T lineage commitment in bone marrow-derived thymus seeding hematopoietic progenitor cells. Prolonged T cell deficiency due to limited thymic function can present a significant challenge in a variety of clinical scenarios. Given the extremely limited availability of thymus transplantation we developed an approach to address this unmet need by engineering an implantable thymic organoid (ITO). Tissue engineering approaches based on implantation of biomimetic three-dimensional (3D) tissue constructs have been used for several decades for *in vivo* regeneration of various tissues and organs. Engineering of functional lymphoid tissues is challenging due to the specific requirements for lymphoid differentiation and has barely been accomplished to date. The aim of our study was to develop a method for engineering of a T cell development supporting 3D microenvironment and to assess if implantation of such a tissue construct is feasible and results in thymus-independent T cell development *in vivo* (Figure 1A).

We designed and tested organoids based on 3D scaffolds fabricated by electrospinning from the FDA-approved biopolymer polycaprolactone (PCL), mouse neonatal thymic tissue fragments containing thymic epithelial cells including epithelial progenitor cells, and functional primary thymic endothelial cells that were engineered with the adenovirus E4ORF1 gene promoting long-term survival while preserving regenerative potential of endothelial cells. Hematopoietic stem and progenitor cells (HSPCs) were seeded to facilitate lymphostromal interactions and sustained functionality both in vitro and in vivo.

Organoid characterization involved laser scanning confocal microscopy for thymic epithelial cell specific markers, live cell imaging, and multicolor flow cytometric analysis of epithelial cells and developing T cell precursors. Our 3D culture system supported growth and expansion of primary thymic epithelial cells for at least 4 weeks (confirmed by live cell microscopy with FoxN1 lineage tracing), in contrast to the historic experience with 2D culture of primary thymic epithelial cells that is not sustainable for more than 24 to 48 hours. Immunofluorescence analysis of thymic organoids revealed that thymic epithelial cell populations included CK8/CK14 double positive thymic epithelial progenitor cells. Mature thymic epithelial cells displayed a predominantly medullary thymic epithelial cell phenotype (CK8–CK14+) (Figure 1B). We optimized the culture conditions by evaluating the impact of the thymic growth factors, bone morphogenetic protein 4 (BMP4) and keratinocyte growth factor (KGF), and the addition of defined numbers of thymic endothelial cells, a cell type that was previously found to boost thymic regeneration. Media supplementation with BMP4 and KGF promoted growth of thymic epithelial cells in vitro. Co-culture of thymic organoids with HSPCs induced their differentiation through successive $CD4^-CD8^-$ (double negative), $CD4^+CD8^+$ (double positive) and $CD4^+CD8^-CD3^+$ or $CD8^+CD4^-CD3^+$ (single positive) stages. T cell receptor (TCR) gene rearrangement is a hallmark of thymic T cell development, and this was confirmed by assessment RAG gene activity in experiments utilizing HSPCs of RAG transcription reporter mice (Figure 1C). The addition of thymic endothelial cells to the organoid further enhanced T cell development in vitro.

In vivo studies included subcutaneous implantation of the optimized thymic organoid systems into athymic mice, along with co-cultured luciferase-transgenic HSPCs. The fate of ITOs was studied via bioluminescence imaging as well as immunophenotyping of lymphoid tissues via flow cytometry, histology and immunohistochemistry to assess both T cell reconstitution and biomaterial responses. Our data indicate that ITOs established *de novo* T cell generation in athymic mice and partially recapitulated thymic function (Figure 1D), and similar to our in vitro findings, expansion of developing T cells within the ITO was enhanced by thymic endothelial cells.

We conclude that thymus-independent *in vivo* T cell generation in a tissue-engineered ITO is feasible, and further development of this technology is expected to produce an exciting innovative strategy for the treatment of T cell deficiency.

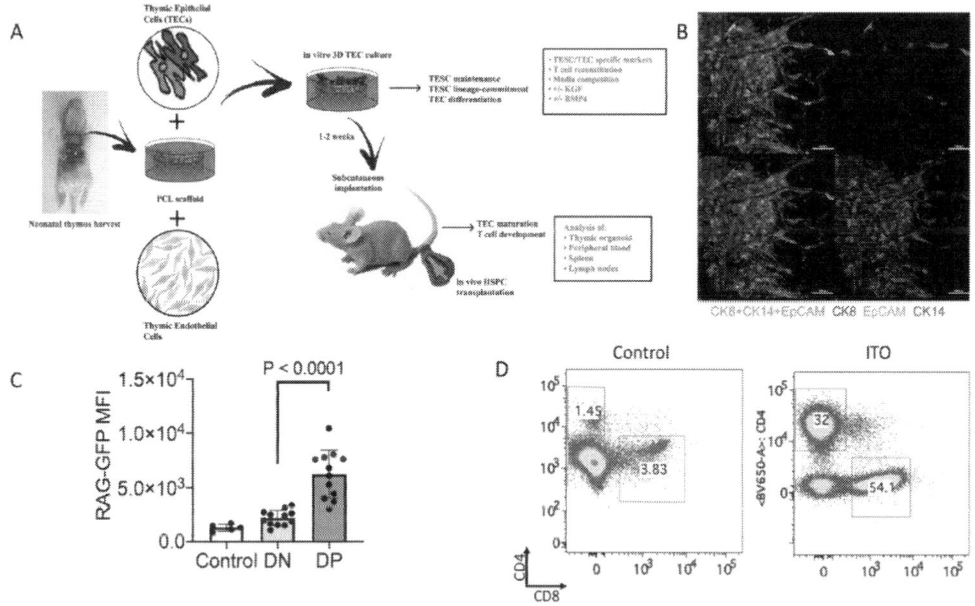

Figure 1. *Establishing T cell immunity with an implantable thymic organoid (ITO).* A) Schematic diagram of the study design. Thymic organoids were engineered from neonatal thymic tissue containing thymic epithelial stem and progenitor cells (TESC), endothelial cells and 3D polymer scaffolds, co-cultured with hematopoietic stem cells (HSPCs) and implanted into athymic nude mice. B) PCL scaffold supports growth of mouse neonatal thymus-derived predominantly CK14-expressing thymic epithelial cells. Mouse neonatal thymic tissue fragments were seeded on a PCL MF/NF scaffold and cultured for 12 days, followed by whole-mount staining and LSCM (n=4). Blue: CK8. Purple: CK14. Green: EpCAM. C) Mouse neonatal thymic tissue fragments were seeded on PCL scaffold and co-cultured in vitro with HSPCs from RAG transcription reporter donor mice. Assessment of RAG gene activity showed TCR gene rearrangement crucial to thymic T cell development. D) Representative FACS plots comparing splenic T cells (CD4+ and CD8+ subsets of CD3+ splenocytes are shown) from control and ITO mice six months post-implantation; consistent with de novo T cell generation in athymic nude mice.

114a Biomaterial Based Approaches to Manage Osteoarthritic Pain

Lakshmi Nair

Pain is considered the most common disabling symptom of osteoarthritis (OA). Currently employed interventions to manage osteoarthritis pain include physiotherapy, anti-inflammatory drugs, intra-articular corticosteroid injections and visco-supplementation such as hyaluronic acid injection. As the prevalence of OA is steadily increasing, there is an urgent need to explore novel therapeutic strategies to manage osteoarthritic pain. The overall goal of our studies is to develop non-addictive approaches to manage OA pain. Our studies on combining biomaterial based approaches with local anesthetics and anti-inflammatory/analgesic pain modulators as well as intraarticular and periarticular routes of administration will be discussed.

114b Ultrasound-Derived Acoustic Radiation Force and Cell Therapy: Combined Technologies for Better Bone Repair

Yusuf Khan

Healing of large-scale bone defects remains an unsolved clinical challenge in the orthopaedic realm. Clinical strategies like autografts, allografts, and bone graft substitutes, and preclinical strategies like tissue engineering, and regenerative medicine all have viability as solutions to this challenge and many have been implemented with varying degrees of success. We have sought to combine both clinical and preclinical strategies into a solution for healing large-scale defects by combining cell delivery via a biocompatible, biodegradable hydrogel with low intensity pulsed ultrasound (LIPUS), a clinically proven method of reducing fracture healing time, into one approach that permits the implantation of cell-laden hydrogels into large-scale bone defects followed by intermittent transdermal application of LIPUS-derived acoustic radiation force (ARF) to the implanted cells. Our work has shown elevated levels of mineralization in hydrogels loaded with marrow-derived stem cells and exposed to ARF over the same cells with no ARF. We have also shown the ability to modulate the extent of mineralization by varying the ultrasound intensity and hydrogel composition. In vivo testing has demonstrated healing of critical size defects when ARF was applied to cell-laden hydrogels, something that was not evident without ARF. Finally our *in silico* modeling allows us to predict the localized forces applied to a cell within the hydrogel for any specific hydrogel stiffness and/or ARF intensity. These preclinical studies bring this novel approach closer to the clinical realm as a novel tool for healing large scale critical size bone defects.

114c Advances in particulate therapy for inflammatory disease

Jamal Lewis

Current paradigms for the treatment of autoimmune diseases (e.g. rheumatoid arthritis [RA]) are woefully inadequate, often missing the mark on desired physiological responses and not targeting the root cause of the disease. Predictably, novel approaches to re-establish immune homeostasis in patients afflicted by autoimmune conditions are now under intense investigation. Notably, we are developing an array of multifunctional, biomaterial-based 'regulatory vaccines' that can be easily administered to remediate some of the prevalent inflammatory diseases. In this talk, I will focus on three particulate systems currently under development in my lab, which attempt to control critical cellular and humoral mediators that engender conditions such as RA, osteoarthritis and autoimmune autism. Additionally, the Lewis lab is currently investigating the interaction of innate immune cells and biodegradable polymers (e.g. PLGA). More specifically, we are interested in deciphering the mechanisms that govern the effects of these materials on innate immune cells.

114d NANOG Induced Skeletal Muscle Reprogramming Enhances Innervation after Peripheral Nerve Injury

Pihu Mehrotra, Shahryar Shahini, Nika Rajabian, Yali Zhang, Jianmin Wang, Song Liu, James Jablonski, Susan Udin, Stelios Andreadis, Kirkwood Personius

Peripheral Nerve Injuries (PNI) affect more than 20 million Americans and severely impact the quality of life by causing long-term disability. These injuries disproportionately affect young individuals and soldiers who fail to regain normal movement and demonstrate persistent weakness. The onset of PNI is characterized by nerve degeneration distal to the nerve injury resulting in long periods of skeletal muscle denervation. Since the speed of axon regeneration post injury is very slow (1-2mm/day), the skeletal muscle becomes atrophied by the time the nerves regenerate and is incapable of "accepting" innervation. To this end, we hypothesize that reprogramming the skeletal muscle to an embryonic-like state may preserve its reinnervation capability following PNI. Previous work in our lab had demonstrated that the transcription factor NANOG can rejuvenate skeletal muscle by increasing the number of $Pax7^+$ muscle progenitors and causes formation of embryonic myosin heavy chain (eMYHC) positive immature muscle fibers in vivo.

To study the effect of NANOG on skeletal muscle innervation, we generated a mouse model (Col1a1-tetO-Nanog, in the ROSA26 locus, hereafter referred to as ROSA-NANOG) in which NANOG expression can be induced upon exposure to doxycycline (Dox). To limit NANOG expression only in the muscle, a slow-release polymer (Elvax), was implanted subcutaneously near the tibialis anterior (TA) muscle for 2 weeks. Elvax implantation was accompanied by sciatic nerve transection and end-to-end repair in both ROSA-NANOG and wild type (WT). WT served as controls. Needle electromyography (EMG) recordings demonstrated that ROSA-NANOG mice elicited almost 3x increase in EMG amplitude as compared to WT mice 16wks post nerve injury, confirming improved reinnervation is a result of re-engineering the muscle to a permissible state. This functional recovery in muscle activity on nerve stimulation accompanied significant improvements in the toe-spread reflex of ROSA-NANOG mice at both 5wks and 16wks post injury. Interestingly, we found that NANOG expression in the muscle prevented atrophy and resulted in increased isometric force production as compared to wild type animals. Further, we performed whole muscle immunohistochemistry to quantify the overlap of muscular acetylcholine receptors (AChRs) with neurofilament and synaptic vesicles. Indeed, while there was very limited presynaptic overlap in WT muscle, ROSA-NANOG mice demonstrated extensive overlap between synaptic vesicles and AChRs indicating restored innervation. In conclusion, we demonstrate that re-engineering the muscle by ectopic NANOG expression can improve functional outcomes after PNI.

114e Polyethylene Glycol Microgel-Based Hydrogels for Neural Stem Cell Growth

Diego Gomez-Maldonado, Narges Yazdani, Rebecca Willits

Our goal, as a research team, is to understand how neural cells interact with their environment to design novel solutions that support nerve regeneration. Neural stem cells have the potential to give birth to new neurons through adulthood, yet little is known about niche microenvironment influences cellular response. As neural stem cells sense both the presence of bioactive signals and the mechanical properties of the extracellular matrix, microgel-based scaffolds are considered as a promising platform with the potential to decouple the mechanical and bioactive characterization of the microenvironment. We hypothesized that this platform could provide integrin-mediated adhesion to alter neural stem cell growth and differentiation through mechanotransduction. In this work, we developed hydrogel systems based on polyethylene glycol diacrylate (PEGDA) microgels generated via precipitation polymerization in an aqueous environment. First, 3 kDa PEGDA was synthesized with a 96% degree of acrylation based on H-NMR measurements. Then, it was combined with different ratios of acryl-PEG-amine to form functionalized microgels serving as building blocks for the hydrogels. These microgels will be characterized for morphology, size, and amine functionalization. Multiarmed PEG-NHS is then used in different concentrations to couple the microgels and modulate the stiffness of the formed scaffolds. The bioactive protein, laminin 511-E8 fragment, was used to promote stem cell adhesion through integrin $\alpha 6\beta 1$. Neural stem cells were cultured on the scaffold to investigate their growth and differentiation and mechanotransduction was measured via YAP translocation. This study will help increase the understanding of complex molecular pathways, which in turn would help develop new tissue engineering alternatives.

114f Acellular and Growth Factor Free Hydrogel Promotes Arteriole Growth and Prevents Tissue Damage in a Murine Model of Critical Limb Ischemia

Corinne Curry, Hayden Paige, Zander Schwartz, Atsuyuki Kawabata, Jeffrey Schmeckpeper, Ethan Lippmann

Statement of purpose: Critical limb ischemia (CLI) occurs when plaque buildup in the arteries becomes so severe, blood cannot reach the extremities, resulting in painful ulcers and non- healing wounds. Patients suffering from CLI often do not qualify for revascularization procedures, leaving amputation as the primary treatment option. This has led to a push for non- invasive treatments that can promote growth of collateral arterioles to restore blood flow and relieve ischemia. Within this space, hydrogels have been employed as carriers for exogenous biologics, such as growth factors and stem cells. However, less research has focused on how the hydrogels themselves can be engineered to promote collateral arteriole development. Here, we present a gelatin hydrogel, modified with a bioactive peptide, as an off-the-shelf vascular therapy that prevented tissue damage in a murine model of CLI. The gelatin hydrogel is functionalized with a peptide derived from the extracellular epitope of Type 1 cadherins, which stimulates growth of smooth muscle cells for arteriole formation in both ex vivo and in vivo assays. The peptide-functionalized gelatin, termed "GelCad," restored perfusion after 14 days in BALB/c mice subjected to a femoral artery ligation. Additionally, preliminary data shows GelCad prevented necrosis in the same femoral artery ligation model applied to 22-month-old C57BL/6 mice. These studies offer a localized, easily translatable vascular therapy that does not require exogenous growth factors or cells.

Methods: The modified gelatin biomaterial is synthesized by dissolving porcine gelatin in PBS and using EDC-NHS chemistry to attach the carboxylate end group of the cadherin peptide (HAVDIGGCE) to primary amines on the gelatin backbone. The peptide attachment is confirmed using H-NMR. For experimental use, the biomaterials are reconstituted as a 10% solution in PBS and crosslinked by mixing with an equal volume of a 10% solution of 20 kDa 4-arm polyethylene glycol succinimidyl glutarate (PEG-SG).

Hydrogels were injected into the fat pad of young C57BL/6 mice. After seven days, the mice were cardiac perfused with Microfil polymer to visualize vasculature in the hydrogel plugs by microCT imaging. MicroCT vascular networks were reconstructed and quantified using Vesselucida software. Separate hydrogels were harvested for cryosectioning and antibody staining to confirm the presence of arterioles.

Femoral artery ligation studies were conducted using 3-month-old BALB/c mice or 22-month-old C57BL/6 mice. In this model, blood flow was restricted by tying double knots in the femoral artery at two locations. Hydrogels were applied on top of the exposed tissue in the subcutaneous space. After surgery, at different time points, laser doppler imaging was performed to assess perfusion in both legs and ischemic indices were used to score physical appearance. At the final endpoint, Microfil was used to cast the vessels for microCT imaging of the vascular network within the hydrogels. Histology was also performed on the fixed tissue by a blinded pathologist to assess ischemic damage.

Results: GelCad hydrogels injected into fat pads showed larger vessel diameters compared to unmodified gelatin hydrogels. Additionally, arterioles were identified by alpha-smooth muscle actin (α-SMA)+ cells lining CD31+ vessels in the GelCad hyrogels that were not present in the gelatin control.

In young BALB/c mice subjected to femoral artery ligation, GelCad hydrogels prevented tissue necrosis and restored perfusion within 14 days, whereas the sham and gelatin controls lost their feet or became completely necrotic after 7 days. Blinded histology analysis confirmed the lack of tissue damage to the GelCad hydrogel-treated mice.

Preliminary experiments in aged C57BL/6 mice subjected to femoral artery ligation have shown full recovery in mice treated with GelCad hydrogels compared to gelatin. Again, the GelCad hydrogels prevented ischemic tissue over 14 days, while the gelatin hydrogel-treated legs were necrotic after 7 days. Unlike the younger mice, full perfusion in the older mice was not achieved, suggesting less collaterals were needed for sufficient blood flow.

Conclusions: Our data suggest that GelCad hydrogels are a promising technology to restore blood flow to ischemic tissues without exogenous growth factors or cells. Future work will focus on rounding out the aged mouse study, optimizing the GelCad hydrogel formulation, and assessing its performance in the presence of other comorbidities.

203a Supramolecular fibrous hydrogels for treating pelvic organ prolapse

Beverly Miller, Monique H. Vaughan, Steven Caliari

Pelvic organ prolapse (POP) is a common pelvic floor disorder that affects approximately 50% of women who have given birth, significantly impacting their quality of life. There is a tremendous clinical need for improved therapeutic approaches to this challenge with the FDA's recent removal of transvaginal mesh kits from the market and the relatively high failure rates of native tissue suture-based repairs. We recently developed an injectable fibrous hyaluronic acid (HA) hydrogel to augment native tissue repairs of POP and tested its efficacy in a small animal model we invented. Supramolecularly-assembled HA hydrogel nanofibers encapsulated in a matrix metalloproteinase (MMP)-degradable HA hydrogel created an injectable scaffold showing excellent biocompatibility and hemocompatibility. The hydrogel could be successfully delivered and localized to the surgery site, where it gradually degraded over 6 weeks. . Together, these results support the potential of this hydrogel platform to augment native tissue POP repair.

203b Effective RNA Delivery using Janus Base Nanopieces against Cartilage Diseases

Yupeng Chen

Janus base nanotubes (JBNts), a novel family of non-covalent nanomaterials, are self-assembled from Janus bases biomimicking DNA base pairs. Due to their versatile and biomimetic nature, we can engineer these nanotubes into a variety of non-covalent structures and devices for biomedical applications. My research focuses on developing JBNts into tiny, rod-shaped RNA delivery vehicles, named Janus base nanopieces (JBNps). We assembled JBNps by "sandwiching" negative charged delivery cargos (such as RNA) among JBNts. Different from conventional spherical vehicles, JBNps have a superior ability to penetrate tissue matrix barriers and deliver RNA cargos into "difficult-for-drug-delivery" tissues, such as articular cartilage and brain. As a proof-of-concept example, we have used JBNps to successfully deliver therapeutic RNA into cartilage for osteoarthritis treatment in a preclinical animal model.

203c Cell-Free Tailored Therapies for Improved Tendon Healing

Brittany Taylor

Despite advances in tendon repair surgery, there is a high re-tear rate due to the inherently poor healing capacity of tendinous tissue. Thus, there has been a shift in paradigm to utilize regenerative medicine to overcome the tendon repair deficiencies and complement the healing response. In particular, cell-free therapies have emerged as an attractive therapeutic to circumvent cellular related complications and simplify the regulatory pathway towards clinical application. To this end, the Taylor Lab investigates the tendon reparative process to inform the development of cell-free strategies, such as sustained therapeutic delivery systems which are designed to specifically address the unique challenges with tendon repair. This talk will focus on the engineering and evaluation of tunable synthetic and biological delivery systems for targeted tendon healing. The innovative design of the synthetic delivery system incorporates nanofibrous matrices that recapitulate the microstructure of native tendon and promote tissue integration,

and depots of biofactor-releasing microspheres to locally regulate cellular activity throughout the tissue healing cascade. Additionally, tailored extracellular vesicles that were engineered using biophysical stimulation will be highlighted. Altogether, this collection of work exploits the capability of transformative technologies to improve tendon repair and broadly contributes to the musculoskeletal field by advancing the translation of cell-free therapies.

203d Dielectrophoretic Label-Free Assessment of Critical Quality Attributes of Therapeutic Extracellular Vesicles Biomanufactured for Regenerative Medicine

Mikhail Skliar, Yunshan Wang

Extracellular vesicles (EVs) are lipid membrane particles inheriting molecular content (e.g., RNAs, proteins) from secreting cells. Lateral signaling by EV uptake was shown to regulate cell senescence, apoptosis, and proliferation, suggesting its therapeutic potential in regenerative medicine [1]. Adult Mesenchymal Stem Cell (MSC)-derived EVs have shown similar efficacy to stem-cell therapy but with improved safety profile due to EVs' small sizes and lack of replication [2]. Therapeutic EVs are produced in clinically meaningful quantities inside bioreactors. However, the biomanufacturing of EVs with consistent properties is an unresolved challenge. The critical impediment is the lack of critical quality attributes (CQA) of EVs, which we can rapidly and continuously assess during biomanufacturing, compare it with nominal CQA of therapeutic EVs, and adjust biomanufacturing conditions of producer cells to maintain the secretion of EVs that match the desired profile.

In this presentation, we will propose a novel dielectrophoretic-based CQA for therapeutic EVs and discuss our progress our progress toward developing tools and methods for its rapid assessment to provide bioreactor feedback to compensate for biomanufacturing disturbances. The screening platform we are developing uses gated insulator-based dielectrophoresis (g-iDEP) to fractionate, concentrate, and the light scattering visualization of EVs' heterogeneity by fractionating and retaining their subpopulations at distinct gates defined by a range of DEP and hydrodynamic forces. Our simulation predicts retention zones where forces on particles with unique combinations of properties (size and conductivity) are balanced, creating a trapping pattern that can be read by light scattering. For example, Fig. 1 (A-D) shows that the increasing conductivity (related to membrane protein concentration and EVs zeta potential) shifts the retention pattern while the size of the retention zone increases above a threshold conductivity. Fig. 1 (E-G) shows that the size of the retention pattern increases with particle sizes, and the two isolated retention patterns merge into one for large particles. Experimentally, we have fabricated a DEP chip and its holder by Nanoscribe 3D Photonic Professional GT2 printer. The retention patterns in the prototype DEP device are interrogated by light scattering. Early computational and experimental results suggest the high sensitivity of the proposed CQA to changes in EVs properties and its suitability to detect the deviation from the nominal retention pattern expected for therapeutic EVs.

[1] De Jong OG, Van Balkom BW, Schiffelers RM, Bouten CV, Verhaar MC. Extracellular vesicles: potential roles in regenerative medicine. Front Immunol. 2014 Dec 3;5:608. doi: 10.3389/fimmu.2014.00608. PMID: 25520717; PMCID: PMC4253973.

[2] Yu X, Huang C, Song B, Xiao Y, Fang M, Feng J, et al. CD4+CD25+ regulatory T cells-derived exosomes prolonged kidney allograft survival in a rat model. Cell Immunol (2013) 285(1–2):62–8. 10.1016/j.cellimm.2013.06.010

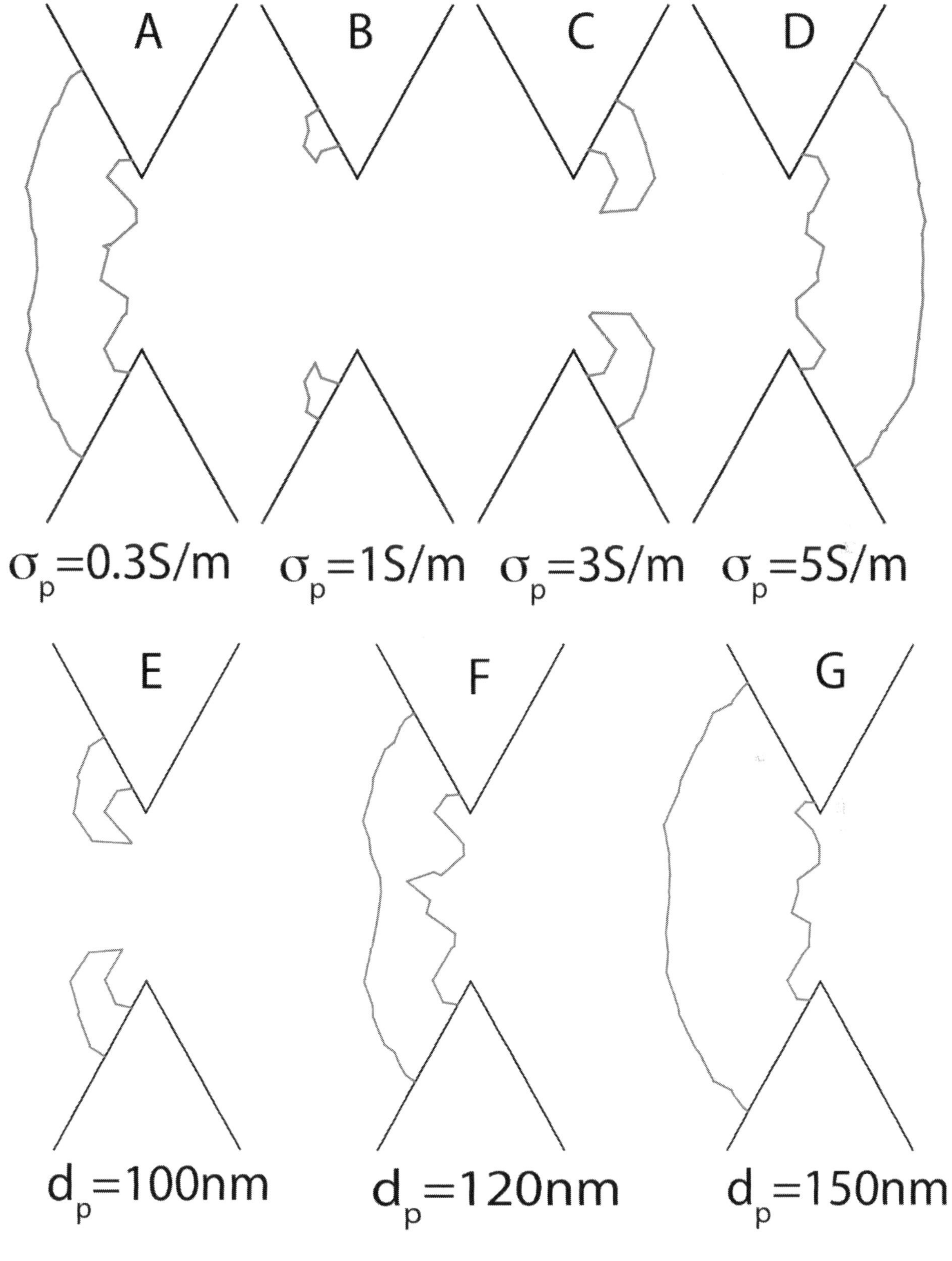
A
B
C
D
σp=0.3S/m
σp=1S/m
σp=3S/m
σp=5S/m
E
F
G
dp=100nm
dp=120nm
dp=150nm

203e Dielectric Characterization of Differentiated Human Stem Cells Using Dielectrophoresis Technique

Raphael Oladokun, Soumya K. Srivastava

Stem cell holds tremendous promise for medical treatments because of their multi-functionality and capacity to self-renew or differentiate into any adult cell type in the body. Undifferentiated stem cells have the potential to develop into cells that serves numerous function and differentiate to yield specialized cell types, which involves switching from proliferation to specialization, i.e., from unspecialized to specialized stem cells under known factors and conditions.

A limited ability to select homogenous populations of stem cells prior to initiating or following a differentiation protocol is a significant challenge for developing effective regenerative tissue engineering therapies and discovering new drugs. A novel way to identify a homogenous or differentiated stem cell population is by utilizing dielectrophoresis, an electrokinetic technique. Our central hypothesis is that the changes in cell membrane capacitance are an electrophysiological property of the plasma membrane that serves as a biomarker for stem cells, transmembrane cell-cell junction proteins, and the cytoskeleton of MSCs with adipogenically differentiation regulate dielectric parameters (conductivity, σ, and permittivity, ε), thus affecting the bioelectric signals of differentiated stem cells, that aid in the identification and selection of a homogenous or differentiated stem cell population affecting the dielectric response of the cells, which can be used for label-free sorting and improved stem cell differentiation process. This hypothesis is based on our preliminary data demonstrating the unique dielectric properties of Skin-derived Stem Cells (SDSCs), Induced Pluripotent Stem Cell (IPSC), and Bone Marrow-derived stromal Stem Cells (BMSCs). We used the experimental crossover frequencies to investigate the behavior of the cells in a medium of known conductivity in the presence of a non-uniform field when cells transduce external signals into intracellular pathways. The short-term objective is to use the dielectrophoresis technique to characterize undifferentiated and differentiated stem cells. The long-term aim is to use the results to develop a label-free, rapid, high-throughput, and low-cost technology platform to select and enrich differentiated stem cells for advanced regenerative therapies.

203f Soft, Resorbable Nanofibrous Mesh for Pelvic Organ Prolapse

Preethi Murugan, Jayesh Bellare

Pelvic organ prolapse (POP) is a medical condition faced by menopausal women wherein pelvic organs like uterus, bladder and rectum descend from their anatomical position. There are four stages or degrees of prolapse as mentioned by POP-Q system, an accepted grading system by multiple professional societies. First degree prolapse can be cured by modifying lifestyle, reducing weight and performing Kegel exercises. But higher degree prolapse requires a mesh that is surgically inserted to give mechanical support of the pelvic organs.

Mechanical support to the pelvic floor can be provided by biological grafts (autografts and allografts, xenografts) and synthetic meshes. Biological grafts have advantages because they support tissue remodelling, but they have disadvantages like immunogenicity (for allografts and xenografts), limited availability, donor site morbidity, disease transmission, mechanical property mismatch, etc. Therefore, there is a need for synthetic meshes that overcomes the disadvantages of biological grafts. To overcome these disadvantages, non-degradable meshes of polypropylene, a synthetic polymer, in the past were widely used by direct implantation to treat POP, but they have come into disrepute recently due to adverse surgical implantation sequelae.

The surgical route is either transabdominal or transvaginal, chosen by the surgeons based on the compartment that got prolapsed. In the past, to treat this condition, non-biodegradable polypropylene hernia meshes were implanted to provide support to pelvic organs, but due to mesh stiffness and mechanical property mismatch, this mesh perforated the neighbouring organs, caused bacterial mesh infection, dyspareunia etc., and compromised women's quality of life. As a result, in April 2019, regulatory agencies like USFDA banned all synthetic meshes that were used to treat POP, from the market.

To overcome all these disadvantages posed by biological grafts and non-biodegradable polypropylene mesh, there is a need for biodegradable, biocompatible, bacteriostatic mesh that mimics the mechanical property and extracellular matrix of the native tissue namely, vaginal tissue. Here we report a novel nanofibrous mesh that was exclusively designed and developed as a mechanical support for the treatment of pelvic organ prolapse.

The properties that need to be considered while designing a nanofibrous biodegradable pelvic mesh are i) biocompatibility, ii) biodegradability, iii) ability to enhance collagen production, iv) mimics the mechanical properties of the vaginal wall (ultimate load, ultimate tensile strength, strain at break and modulus) and v) bacteriostatic ability.

To obtain the above property we chose biodegradable polymers along with additive. The polymers we used to design our biodegradable mesh are polycaprolactone (PCL) and citric acid functionalised polyethylene glycol (PEGC). PCL and polyethylene glycol (PEG) are FDA approved biocompatible and biodegradable polymer. PCL is used for its ease of processibility, good tissue integration, negligible inflammation and slow degradation, which gives enough time for the tissues to laydown extracellular matrix (ECM). PEGC is prepared by functionalising citric acid with PEG through thermal polycondensation reaction. Citric acid is a multifunctional acid, metabolic product of Krebs cycle, has a role in cell proliferation and metabolism. Citric acid is known to promote collagen production *in-vivo*, therefore we hypothesis that functionalisation of PEG with citric acid to give PEGC will enhance the collagen production, which is essential for pelvic floor repair and regeneration. PEGC addition in the mesh formulation imparts elasticity to the mesh, improves hydrophilicity and tunes mesh degradation. To this formulation zinc oxide nanoparticles (ZnO) are added to improve mechanical properties (ultimate load, ultimate tensile strength, strain at break and modulus) and collagen production. ZnO also imparts antibacterial property whose addition into mesh imparts bacteriostatic ability.

The polymer, additive and nanoparticles are electrospun into nanofibrous mesh. The composition of PCL-

PEGC-xZnO meshes produced through electrospinning are 90%PCL, 10%PEGC and ZnO (x=0, 0.1, 0.5 and 1 wt% ZnO w.r.t PCL). With increasing ZnO loading into the nanofiber, the fiber diameter of mesh increased from 495 nm to 900 nm, which improved the strain at break of the mesh. The design criteria for mechanical property of mesh is chosen based on the mechanical property of the vaginal wall: the minimum ultimate load- 1N, minimum ultimate tensile strength- 1 MPa, minimum strain at break- 0.7 mm/mm and modulus range 5 to 11 MPa. The mechanical properties of mesh even after degradation up to 28 days and terminal sterilization using gamma irradiation was similar to that of vaginal wall. The mechanical properties of the commercial polypropylene mesh were also measured, and it was observed that their mechanical properties were not similar to that of a vaginal wall: the ultimate load is 102x higher, ultimate tensile strength is 29x higher, modulus is 7x times higher and strain at break is 35x times lower than the vaginal tissue which is not desirable as a mesh for pelvic organ prolapse, whereas our mesh perfectly matched the mechanical properties of vaginal wall.

The biological mesh performance *in-vitro* was tested by using L929, mouse fibroblast cells. The mesh formulation PCL-PEGC-x ZnO (x=0, 0.1, 0.5 and 1 wt% ZnO w.r.t PCL) was tested for biocompatibility, collagen secretion by L929 and the effect of ECM secretion on the mechanical properties of the nanofibrous mesh. The formulations PCL-PEGC and PCL-PEGC-0.1 ZnO mesh were biocompatible, haemocompatible and enhanced the collagen production *in-vitro* in comparison with commercial mesh. Out of which PCL-PEGC-0.1 ZnO mesh supported the highest collagen production. Mesh formulations containing high loading of ZnO like PCL-PEGC-0.5 ZnO and PCL-PEGC-1 ZnO were found to be not favourable for L929 growth due to higher ZnO loading. Therefore, PCL-PEGC-0.1 ZnO was chosen as the optimum formulation.

When deployed, the mesh would be infiltrated with fibroblast that would laydown ECM. Effect of ECM on the mechanical properties of the meshes were studied by culturing the L929 on the meshes for a period of 28 days. The collagen deposition on the PCL-PEGC-x ZnO (x=0, 0.1 wt% ZnO w.r.t PCL) improved their mechanical properties (ultimate load, ultimate tensile strength, strain at break and modulus) and their values were in the correct range of the mechanical design criteria for the mesh.

The meshes were evaluated for their bacteriostatic ability *in-vitro* using *E. coli* and *S. aureus*. In comparison to the commercial mesh, PCL-PEGC-x ZnO (x=0, 0.1, 0.5 and 1 wt% ZnO w.r.t PCL) meshes exhibited 2 log reduction in bacterial adhesion.

Based on extensive testing and analysis presented above, the nanofibrous mesh that we formulated and fabricated possessed all the criteria such as softness, elasticity, proper mechanical property and ability to promote ECM production to give a natural fibrous support membrane as the scaffold mesh degrades, all of which are essential for its application in pelvic organ prolapse repair, making it an appropriate alternative solution for the USFDA banned non-degradable polypropylene mesh. With these better matched properties the mesh developed here can be taken up for preclinical studies.

Soft, Resorbable Nanofibrous Mesh For Pelvic Organ Prolapse

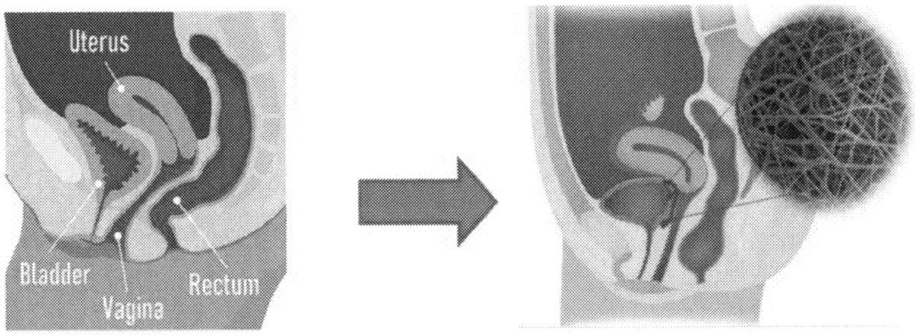

- Biocompatible
- Biodegradable
- Bacteriostatic
- Collagen production
- Mimics mechanical property of vaginal wall

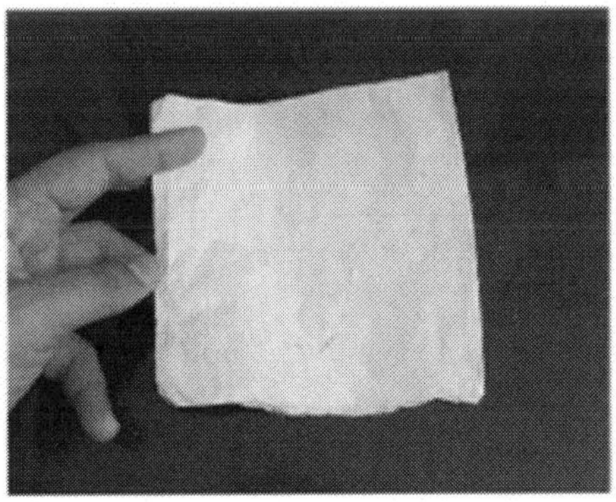

Effect of ECM on the mechanical properties of nanofibrous mesh

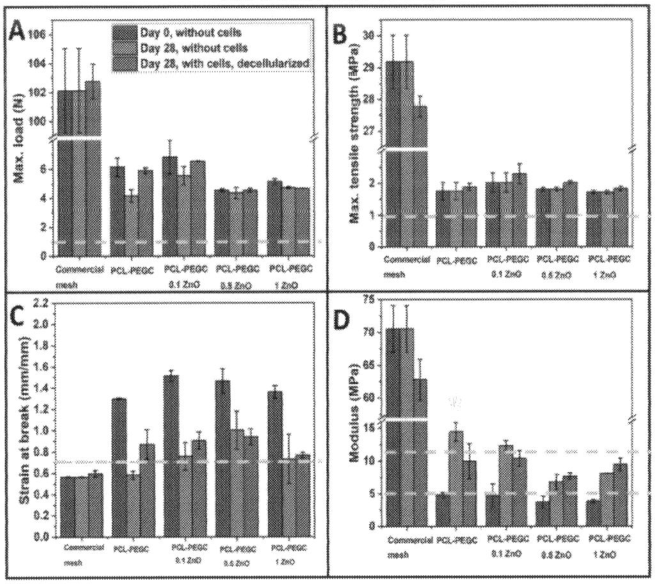

AUTHOR INDEX